I0464611

.

NCA Report Series, Volume 9: Uses of Vulnerability Assessments for the National Climate Assessment

NCA Report Series

The National Climate Assessment (NCA) Report Series summarizes regional, sectoral, and process-related workshops and discussions being held as part of the Third NCA process.

The workshop on including and conducting vulnerability assessments as a part of the 2013 NCA was held in Atlanta, Georgia on January 19-20, 2011. Volume 9 of the NCA Report Series summarizes the discussions and outcomes of this workshop. A list of completed and planned reports in the NCA Report Series can be found online at http://assessment.globalchange.gov.

CONTENTS

I. Executive Summary

The NCA Vulnerability Assessment Workshop, co-sponsored by the U.S. Department of Agriculture Forest Service and the Centers for Disease Control and Prevention, was held on January 19-20, 2011 at the Embassy Suites Atlanta-Buckhead Hotel in Atlanta, Georgia. A group of 69 experts (see list in Appendix B) convened to share their experiences. Participants brought a wide range of disciplinary expertise in the natural and social sciences, sector expertise, and experience with the concepts of vulnerability and conducting vulnerability assessments (VAs). Participants included representatives from federal and state government, non-governmental organizations, tribes, universities, and communities. This workshop was the first engagement with the NCA for many participants, allowing for productive dialogues on connecting to new networks and knowledge-sharing.

The workshop participants were asked to provide input on a number of topics, including guidelines and criteria for VAs to be used in the 2013 NCA report and future NCA processes and products; the relevance of existing VAs to the NCA; how VAs can incorporate multiple stressors and social/economic factors that can determine sensitivity and are relevant to coping with hazards and climate threats; and how to assess the adaptive capacity of natural and built environments.

Participants' input was received through in-depth discussion in breakout sessions, plenary sessions on break-out results, and several panels that provided key insights about VAs, lessons learned through experience with applied VAs, and reflections on group discussions (see Agenda in Appendix A).

Breakout groups were charged with addressing questions that generally fell within four main themes over the two-day workshop: (1) Current state of VAs; (2) Criteria for evaluating existing VAs; (3) VA framework for the 2013 NCA Report; and (4) A sustainable process for future NCA VAs.

1.   Current State of Vulnerability Assessments: Informing the NCA

Workshop participants were asked to evaluate existing VAs that could be used in the 2013 NCA. They identified overall strengths and weaknesses based on existing literature and other sources. The group identified more weaknesses than strengths, which likely reflects the relative "newness" of VAs in the context of climate change and variety of disciplinary approaches and lack of agreement among them. Participants discussed that the strengths of VAs include some sectors having a wealth of existing assessment information and progress that has been made in linking climate issues to decision-making processes. Additionally, existing assessments are increasingly using modeling, mapping and geographic information systems (GIS) to improve analysis and visualization of vulnerabilities.

One of the weaknesses of VAs is that the word "vulnerability" is used in several different ways within and across disciplines. Moreover, the communication between scientists and the communities experiencing impacts should be improved. The data and information needed to do VAs is often incomplete or inaccessible and slow-onset variables that gradually alter individuals' or systems' vulnerability are not well understood. Also, there is an over-reliance on climate models, which limits the ability of those doing assessments to quantitatively include some key risks/exposures that affect vulnerability and are not accounted for in climate models. Additionally, the interaction between climate change stressors and other stressors needs to be better understood and the scientific peer-reviewed literature on VAs is growing, but is still sparse. VAs tend to focus more on negative outcomes, when there are also important positive outcomes to document as well. Also, more experience with doing VAs at regional or higher levels is needed to make linkages to the decision-making processes.

Participants raised many challenges that the NCA will face in improving VAs. One of the challenges is making VAs more policy or decision relevant; using VAs as a tool to communicate with decision-makers could help overcome institutional barriers to adaptation. Another challenge is conducting VAs with different levels of resource input to better understand the value added by increased resource input; combining public and private resources may be an important strategy for leveraging the resources for VAs to support climate change adaption needs. More work needs to be done to better understand the challenges of scaling-up and stitching together place-specific vulnerability studies for generalized results relevant to an entire sector or region. Additionally, defining system boundaries poses a significant challenge; climate impacts can be localized, but the parties that need to be involved in responding to the impacts are often more widely distributed. Many of these challenges will not be addressed by 2013 due to resource and time constraints, but should be considered for future assessments.

2. Criteria for Evaluating Existing Vulnerability Assessments

Workshop participants developed a general set of criteria for evaluating VAs for use in the 2013 NCA, building from a strawman list presented at the workshop. This section summarizes criteria for structural, content and communication components of a VA.

One criterion for structural components of a VA includes indicating if an assessment is determining the vulnerability of a population, geographic area, sector or multiple sectors or some combination of these factors. Also, spatial and temporal scales should be appropriate to the type of vulnerability and particular sector. Vulnerability assessments should consider the changing physical, social and economic conditions in addition to projected changes in climate. Additionally, the review process does not determine the utility of information from a decision-making perspective; the gray literature is a great source for assessments that could be included in the NCA.

The content components' criteria include utilizing both qualitative and quantitative information, using data appropriate for the vulnerability and scale being addressed and making use of the most up-to-date climate and associated impact models when available and/or relevant. Also, effective use of community baseline information is necessary to determine current and future adaptive capacity and uncertainty should be either measured quantitatively or expressed qualitatively. As social, ecological, and economic factors can interact in important ways, they should be considered for all sectors but may vary in degree of detail. Many impacts of climate may manifest themselves as interactions with other key stressors, which the 2013 NCA should begin to identify.

Regarding communication components, local context and knowledge is essential for many decision-making processes and should be included as input into the VA. The use of mapping to describe current states and potential futures can be a valuable tool in promoting understanding and community participation. Assessments should be useful to end-users in making more informed decisions.

3. Vulnerability Assessment Framework for the 2013 NCA Report

Workshop participants were charged with providing input on what an integrated VA might look like in the regional and sectoral structure of the 2013 NCA Report. The group considered new approaches that would be efficient (built from existing assessments), dynamic (connected to updated NCA climate science, modeling, impacts, *etc.*), and most importantly, useful (relevant to current decision-making, especially at local levels).

For the 2013 NCA, vulnerability information should be packaged and communicated in a manner that "leads" with the issues of concern regarding climate impacts on already-stressed or sensitive resources and communities. When issues of vulnerability are raised, there should be practical solutions to match the vulnerabilities. As a major component of VAs, adaptive capacity will need to be addressed in some coherent and consistent manner throughout the 2013 NCA. At a minimum, adaptive capacity needs to be defined (using a common starting definition) for the NCA sectors and cross-cutting issues below. If possible, through the use of existing VAs and expert knowledge, potential "indicators" of adaptive capacity should be identified and used to help frame the VA narrative in the NCA report.

Workshop participants identified the need to take an integrated and iterative approach to VAs in the NCA that includes common linkages across sectors and geographies, considers vulnerability from a holistic "systems" perspective wherever possible and is adaptive as more successful VAs are completed. Some key considerations for this approach include linking to scenarios of possible futures that take other factors besides climate change into account, which could help with integration within sectors and regions, and across the entire assessment, and building on existing climate impact assessments. A critical first step is to conduct a full review in collaboration with the regional and sectoral working groups to determine what climate impact assessments are available and most relevant for use in VAs.

Some of the key activities that need to be considered in moving forward with the VA process for the 2013 NCA include developing the following: a matrix listing targeted "decision-drivers" to help guide the VA process towards informing decision-making processes at national, regional and local levels; a lexicon for use in the NCA process, which will provide reference definitions for the language to be used in the NCA and will serve as a resource for communicating across disciplines in the development and integration of VA information; a hazards/stressors taxonomy to identify and categorize hazards, threats, and environmental stressors in a manner that could support the systematic evaluation of current risks, anticipated climate change impacts, and potential effects of societal changes; and an action-focused communication strategy. It is also important to include case studies grounded in impacts at the local scale and to expand the scope of literature reviews to cross disciplinary boundaries because VAs operate at the intersection of science and policy.

4. A Sustainable Process for Future NCA Vulnerability Assessments

One of the goals of the workshop was to identify ideal core elements of an integrated assessment that could serve as an implementation framework for future NCAs.

Participants envisioned the VA component of the NCA as an iterative, evolving process so that each NCA would learn from the previous NCA, particularly about the Nation's vulnerability to climate change. The NCA would act as a roadmap to provide guidance on the methods of conducting regional and local VAs, especially to ensure that they are cross-sector and include multiple stresses, but also how those assessments would provide information on the national vulnerabilities. This latter aspect might include identifying higher-order questions that link on-the-ground issues with the adaptive capacity of federal institutions. Vulnerability assessments would take the longer-term view, even though the reporting period for the NCA is every four years, and could periodically focus on issue- or theme-based studies to facilitate scaling-up the vulnerabilities. Primarily, the participants suggested that the sustainable process for future NCA vulnerability assessments could be a civic discovery process where engaged stakeholders facilitated the broader learning and understanding about vulnerability in communities across the U.S.

5. Workshop Perspectives

Three participants provided perspectives at the end of each day, highlighting that the NCA offers a major opportunity to build a national conversation about adaptation and mitigation and can be used as a teachable moment. If the NCA is going to provide guidance for VAs, it needs to be dynamic, flexible

and cross-sectoral. Putting long-term processes into place to provide ongoing exchange of information and reaching out to a broader community will be critical to the NCA's success. VAs provide an excellent tool for engagement to lead with what is vulnerable from climate change and the possible futures, rather than start with the climate science. VAs can serve a role in broader conversations about climate change by focusing on what is going on where people live and the environment they are familiar with, which may already be experiencing impacts due to climate change.

Effective communication that encourages stakeholders' engagement in the NCA and helps to build a community of stakeholders is key to the process. This includes reaching out to those who are motivated to be involved in the NCA and associated VAs, such as tribes. A challenge for the NCA is determining the frame of reference and scale for providing information to decision-makers at all levels. This also includes the importance of distinguishing what is feasible for the 2013 NCA report versus the long-term process for planning work and resource allocation, as well as deciding if VAs should be conducted at the national scale or on a regional and/or sectoral scale that can be aggregated.

6. Workshop Conclusions

Developing a **system-based approach** that connects regions, sectors, and multiple stressors in a more cohesive picture of the effects of climate change on human and natural systems will help the 2013 process of stitching together the extant information and facilitate the evolution of the sustaining NCA process of VAs. The systems approach would benefit from coordination across the working groups to identify an agreed upon lexicon, including definitions, as well as the suite of climate-economic-social scenarios and how uncertainty will be described.

The importance of **integrated assessments** was also clear. Understanding the interactions between stressors is critical, as is looking across sectors. Vulnerability assessments need to look more broadly than a particular sector or topic. The emphasis on cross-cutting themes in the planning for the NCA reflects this increased emphasis on integrated analysis, while realizing the challenges of conducting such analyses.

For the 2013 NCA, preliminary work to develop **links to decision-making** would ensure the successful application of information from the VA at the regional and local levels. Local context and knowledge is essential for many decision-making processes and should be included as input into the NCA vulnerability assessment. The 2013 NCA should include both an **action-focused communication** strategy that encourages stakeholders' engagement in the NCA and helps to build a community of stakeholders and a bottom-up approach to incorporate community input and participation, providing communities with a sense of empowerment and ownership of the NCA process.

The importance of equity, environmental justice, and institutions are also dominant in any discussion about vulnerability. Understanding the potential effects of climate change on people and natural systems is essential to making fair and informed decisions about adaptation options. It was emphasized throughout the workshop that often the most vulnerable human populations are often difficult to identify and usually have little political power. Therefore, a fine-toothed knowledge of the social structure, the history, and the demographic composition and its "patchiness" of the population vulnerable to climate change impacts is essential to depict in a sound VA.

Assessing vulnerability in the NCA could be an **ongoing adaptation and learning process** that engages many sectors, regions, and institutions in discussions about vulnerability to climate change and potential adaptation options. Such a process could operate as a continuum providing guidance on approaches for assessing vulnerability to current and future impacts, providing guidance on the qualitative and quantitative methods to assess that vulnerability and a process to synthesize the results into guidance about the risks, and options for mitigation and adaptation.

II. Introduction

National climate assessments act as a status report on climate change science and impacts. They are based on observations made across the country and compare these observations to projections from climate system models. The NCA aims to incorporate advances in the understanding of climate science into larger social, ecological, and policy systems, and with this provide integrated analyses of impacts and vulnerability. The NCA will help evaluate the effectiveness of our mitigation and adaptation activities and identify economic opportunities that arise

as the climate changes. It will also serve to integrate scientific information from multiple sources and highlight key findings and significant gaps in our knowledge. The NCA aims to help the federal government prioritize climate science investments, and in doing so will help to provide the science that can be used by communities around our nation try to create a more sustainable and environmentally-sound plans for our future. The previous Assessment (*Global Climate Change Impacts in the United States*) produced a report that was completed in 2009, and the first National Assessment was completed in 2000.

The 2009 NCA, *Global Change Impacts in the United States*, provided an overview of observed and projected climate change impacts in the United States. The report highlighted a number of risks associated with climate change, including infrastructure risk from hurricanes and wildfires, risks to agricultural production, water resource shortages, sea-level rise and shifts in ecosystems. The scientific consensus from the 2009 report was that climate change will have lasting and wide-spread direct and indirect impacts on human and natural systems in the United States. Despite considerable uncertainty about the timing and magnitude of impacts, the scientific information highlighted in the 2009 report is sufficient to warrant evaluation of those risks using vulnerability assessments. Results from vulnerability assessments can be used to inform decision-making processes related to climate change, including developing mitigation and adaptation plans.

The next NCA will be delivered in 2013. An interagency task force has been planning the process for completing a 2013 NCA report and developing a sustainable process for subsequent NCAs. A federal advisory committee, formally known as the National Climate Assessment Development and Advisory Committee (NCADAC), has been formed to provide advice and recommendations toward the development of an ongoing, sustainable national assessment of global change impacts and adaptation and mitigation strategies for the U.S. The initial planning for the NCA included conducting a series of process and methodological workshops to gather input from experts on a variety of topics. One of the topics identified as important enough to warrant a workshop was the role of vulnerability assessments (VAs) in the NCA.

The NCA Vulnerability Assessment Workshop, co-sponsored by the U.S. Department of Agriculture

Forest Service and the Centers for Disease Control and Prevention, was held on January 19-20, 2011 at the Embassy Suites Atlanta-Buckhead Hotel in Atlanta, Georgia. A group of 69 experts (see list in Appendix B) convened to share their experiences. Participants brought a wide range of disciplinary expertise in the natural and social sciences, sector expertise, and experience with the concepts of vulnerability and conducting vulnerability assessments. Participants included representatives from federal and state government, non-governmental organizations, tribes, universities, and communities. This workshop was the first engagement with the NCA for many participants, allowing for productive dialogues on connecting to new networks and knowledge-sharing.

The workshop participants were asked to provide input on a number of topics, including
- Guidelines and criteria for VAs to be used in the 2013 NCA report and future NCA processes and products.
- The relevance of existing VAs to the NCA.
- How VAs can incorporate multiple stressors and social/economic factors that can determine sensitivity and are relevant to coping with hazards and climate threats.
- How to assess the adaptive capacity of natural and built environments.

Input from the participants was received through in-depth discussion in breakout sessions, plenary sessions on break-out results, and several panels that provided key insights about vulnerability assessments, lessons learned through on-the-ground experience with applied vulnerability assessments, and reflections on group discussions (see Agenda in Appendix A).

To provide context for workshop discussions about VAs within the context of the NCA, a white paper (Mills *et al.*, 2011) was provided in advance of the workshop as an information resource. The white paper reviewed definitions of vulnerability and recent literature on vulnerability assessments. It included the widely accepted concept of vulnerability to climate change as a function of a system's exposure, sensitivity, and adaptive capacity (Figure 1), which were defined as:

- **Exposure** – in the context of vulnerability to climate change, the construct of exposure is used to describe the climate-related stressors that influence particular systems. This can include stressors such as drought (*e.g.*, in the context of water resources, agriculture, forestry) or sea-level rise (*e.g.*, coastal flooding, habitat loss).

- **Sensitivity** – defined as "the degree to which a system is modified or affected by (climate) perturbations" (Adger, 2006). Sensitivity is a measure of how responsive a particular sector or receptor is to climate variability and change. As an example, although all crops in a region might be subject to the same exposure due to a prolonged drought, some crops may be more affected than others due to a higher sensitivity to water-limited conditions. As with the other constructs considered, the definition of sensitivity varies within and across disciplines. For example, in health, sensitivity is often considered to be an individual's or subpopulation's responsiveness to an exposure, primarily for biological reasons (Balbus and Malina, 2009).

- **Adaptive capacity** – this is a measure of a sector's ability to reduce impacts through constructive change. Adaptation can be either "planned" or "autonomous" (Downing and Patwardhan, 2004). Planned adaptation can be thought of as proactive (*e.g.*, a farmer transitioning to more drought-resistant plants before an event), while autonomous adaptation can be thought of as more reactive (*e.g.*, the same farmer switching to more heat-tolerant crops after the climate has warmed). Social systems have the ability to conduct "planned" adaptation through proactive steps; however, natural systems do not.

The white paper also explained that assessing vulnerability requires knowledge about how social and natural systems are influenced by external stressors.

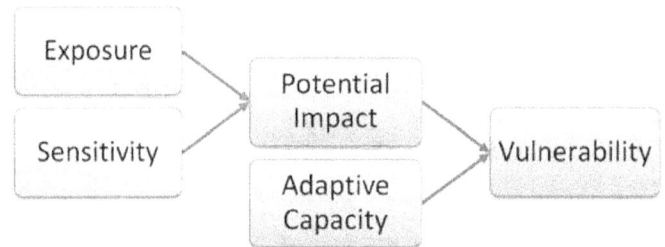

Figure 1. Many sectors and disciplines generally perceive vulnerability as a function of exposure, sensitivity, and adaptive capacity, but definitions can vary considerably.

Practitioners use VAs to forecast a series of future stressors and related responses to these and other external stressors. As VAs are conducted across multiple disciplines, there are many differences across disciplines in data requirements and frameworks for assessment. The specific questions a particular VA is trying to answer will influence the framework used (see general characteristics for framing VAs for each of the NCA sectors in Appendix D).

Charge to the Breakout Groups
Breakout groups were charged with addressing questions that generally fell within these main themes over the two-day workshop:

1. The current state of vulnerability assessments.
2. Criteria for evaluating existing vulnerability assessments.
3. A vulnerability assessment framework for the 2013 NCA report.
4. A sustainable process for future NCA vulnerability assessments.

Common issues and themes arose from all breakout sessions, so outcomes are described thematically rather than chronologically. Some concepts were consistent among the three main areas, and are thus repeated throughout the document.

Throughout the two days, participants tended to discuss vulnerability within the context of their specific discipline or sector. A major point that arose throughout the discussions was the need to clearly define terms and be consistent across groups when using terminology, especially the concept of vulnerability. Participants also tended to go back and forth between discussing the "concept" of vulnerability versus the content of VAs. Many of the participants had extensive experience with vulnerability concepts and frameworks, but not necessarily in the context of the NCA. One of the NCA's challenges is to now translate the concept of vulnerability and local-scale-assessment into a national, integrated framework for the 2013 NCA, and facilitate the evolution of that framework in the sustained NCA process.

The statements in the following sections do not represent consensus of all participants, but are general themes that emerged from individual comments regarding vulnerability assessments during the workshop.

III. Current State of Vulnerability Assessments: Informing the NCA

Workshop participants were asked to identify and evaluate existing VAs that could be used in the 2013 NCA. They identified overall strengths and weaknesses based on existing literature and other sources. The group identified more weaknesses than strengths, which likely reflects the relative "newness" of climate change vulnerability assessments. Although there is a significant body of peer-reviewed literature on the concept of vulnerability and vulnerability assessment frameworks, much of this is conceptual. By contrast, much of the operational VAs are published in the gray literature or remain unpublished.

a) Strengths of Vulnerability Assessments

Some sectors have a wealth of existing assessment information. For example, there is a large body of work in health risk assessments and the links between health variables (*e.g.*, water, air quality, heat) and climate change are well-understood.

Progress has been made in linking climate issues to decision-making processes. Some regions and sectors have used advisory and stakeholder groups to help identify specific decision-making processes affected by climate change. These linkages can help define and target specific vulnerabilities to assess.

Existing assessments are increasingly using modeling, mapping and geographic information systems (GIS) to improve analysis and visualization of vulnerabilities. These tools provide decision-makers with additional options and opportunities for incorporating long-term consideration of risk into their ongoing planning and investment decisions (see Figure 2 on page 12 for a case study using a GIS framework). Mapping improves the ability of the VA to communicate a message to all kinds of people.

b) Weaknesses of Vulnerability Assessments

Lack of clear definitions
The word "vulnerability" is used in several different ways. For example, the disaster anthropology perspective focuses on a social group's vulnerability as a function of their social construct – including characteristics such as income level, race, ethnicity, health, language, literacy, land-use patterns that all affect both individual and community vulnerability. In contrast, the natural resource perspective focuses

on the sensitivity of natural resources to climate and other stressors and the corresponding implied impact on humans from the resource effects. The public health perspective focuses on prevention of adverse outcomes and preparedness (instead of hazard mitigation and adaptation).

Decision-makers are much more familiar with impact assessments and they sometimes presume VAs are similar. Impact assessments are focused on exposure, and do not typically address sensitivity or adaptive capacity. VAs, in contrast, should also address sensitivity and adaptive capacity, which, for sensitivity, require a greater understanding of the biological, physical and social dynamics, and for adaptive capacity, more stakeholder input.

The differences between coping, adaptation, and adaptive capacity are not always well discerned. Although the three can be related, they are separate issues that need to be parsed out to be more effective in producing the desired results. For example, damming of the Mississippi River prevents sediment from reaching the Mississippi River Delta, obstructing the restorative process that built up the coastal land over time. The loss of sediment coming down the river results in a loss of adaptive capacity to floods and storm surges for the people down-stream. Coping to sea-level rise might include re-diverting

sediment into wetlands for a set near-term change in sea level. In contrast, an adaptation response to a stressor such as sea-level rise might be to explore options that would allow some sediment to naturally return as well as planning for increasing rises in sea level.

Incomplete data and information
Application of conceptual frameworks to local assessments is often limited by data availability. The "ideal" data inputs for a VA are often unavailable at the local scale. For example, some types of demographic and economic data are consistently available at the county level, while data at finer scales are often missing or inaccessible. Data on social and economic variables not collected by Census or other national entities are generally unavailable. Similar issues exist for ecological data at fine spatial scales.

Some types of data are inaccessible for particular sectors. It is difficult for researchers to obtain some types of health, social, and economic data that are needed for informed decision-making. For example, privatization of health data is a huge barrier and impairing the ability to obtain data that might be used to assess the vulnerabilities within public health.

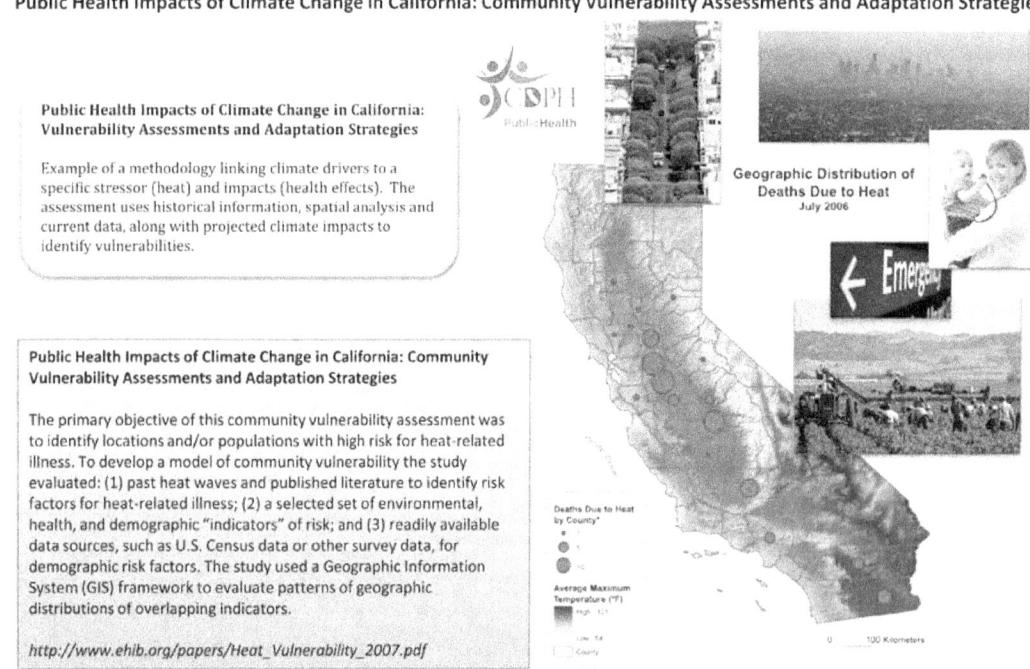

Figure 2. Public Health Impacts of Climate Change in California: Community Vulnerability Assessments and Adaptation Strategies

Shishmaref, Alaska, community needing to relocate due to decreased Arctic sea ice, thawing permafrost, extreme weather events and sea-level rise. ©Tony Weyiouanna, Shishmaref.

Climate models can provide some of the information needed about stressors, but not all of it. Vulnerability assessments rely on existing scientific information and climate models. Climate models provide limited information on extremes in climate or episodic events such as floods, ice storms, or wind events. In addition, future climates will be influenced by and associated with other major driving forces, such as population growth, economic growth, land-use change, and expansion of invasive species. This type of information and modeling is often considered at the end of an assessment and given minimal resources for development. Such information is critical to establishing adaptive capacity as well as identifying adaptation options for natural resources, humans, and communities.

"Slow-onset variables" – that gradually alter vulnerability of affected individuals or systems – are not well understood. Assessments need to be more dynamic over time and consider long-term vulnerability from "slow-onset variables," such as drought and sea-level rise, to provide more than the snapshot view.

Interactions and linkages are weakly described
The interaction between climate change stressors and other multiple stressors needs to be better understood. There is a limited understanding of how long-term non-climatic and climatic stressors interact and how these interactions can increase vulnerability to acute events, such as flooding. For example, people could be driven out of areas by economic and social forces and pushed into zones that are at risk of intensifying hurricanes. This type of understanding is critical for the 2013 NCA.

VAs tend to focus more on negative outcomes, when there are important positive outcomes to document as well. Climate change effects are often presented as grim or dire consequences. Yet, there

are often multiple benefits of mitigation and adaptation policies and actions. For example, urban heat islands affect economically and socially vulnerable communities by increasing energy costs, air pollution and heat-related illnesses and mortality. Installing green roofs through urban heat island reduction programs mitigates greenhouse gas emissions, reduces air temperature, and improves air quality. This also benefits economically vulnerable populations who are at-risk of heat-related impacts, which are predicted to increase and intensify due to climate change.

More experience with doing VA's at regional or higher levels is needed. Most vulnerability assessments have been conducted at local scales to address local needs. Regional and local factors can interact to affect the vulnerability of people, natural systems, or infrastructure to climate change, but greater understanding of these interactions is needed. For example, local topography could make a community more susceptible to flooding than the surrounding region. That community's susceptibility to flooding could also be partly a result of historical policies at a national or regional scale that pushed them onto a floodplain in the first place. The complexity of these interactions has hindered the ability to conduct VAs at higher spatial scales.

Communication barriers
Communication between scientists and the communities experiencing impacts should be improved. Oftentimes, communication between scientists developing climate change and climate impact models and end-users is lacking. Scientific information about climate change is often difficult to translate to the community level where the impacts are most visibly experienced. Scientists developing models to assess climate change impacts often do not get input from end-users in designing the delivery of results. Once model results become available, many communities find it difficult to use information about the effects of climate (*e.g.*, on health) in local decision-making processes. However, programs such as Sea Grant have extensive experience in serving as "science liaisons"; they understand the communicative challenges and have proven strategies to overcome these obstacles.

Public perceptions of risk are important to vulnerability assessments, but incorporating them effectively is still a challenge. Individual perceptions of risk based on people's knowledge, attitudes, and beliefs significantly affect the success of adaptation and mitigation policies and actions. Designing responses to climate or other vulnerabilities without understanding these perceptions could result in less effective outcomes.

Literature gaps
The scientific peer-reviewed literature on vulnerability assessments is growing, but is still sparse. Peer-reviewed literature tends to have a long time-lag embedded in the process and be narrowly focused on particular scientific questions that often do not link to decision-making processes. Private sector analyses or adaptation activities and community-based projects are often not documented in the traditional scientific process. A number of VAs have been conducted and reported in the gray literature that can be highly valuable, as long as the NCA adopts a review process that maintains scientific credibility.

Social science is underrepresented in existing assessments. People and institutions are often overlooked as critical factors in vulnerability assessments. Evaluations of adaptive capacity should include considerations of environmental justice, governance, institutional effectiveness, social networks, community engagement, and risk perception.

c) Challenges in Improving Vulnerability Assessments

Participants raised many challenges that the NCA will face in improving VAs. Some are relevant to the broader scope of the NCA, beyond just VAs. Many of these challenges will not be addressed by 2013 due to resource and time constraints, but should be considered for future assessments.

Making VAs more policy- or decision-relevant. There is value in tying VAs to policy decisions, so that they can be policy-relevant, and tailored, on the outset, to address policy questions. The challenge for the NCA is to facilitate information sharing at the local level, where most adaptation decisions have to be made. In addition, the tendency is to focus on public decision-makers, when private decisions may be influencing vulnerability most directly. Well-mapped social networks are needed to understand who is shaping the decision-making process. Using VAs as a tool to communicate with decision-makers could help overcome institutional barriers to adaptation.

Conducting VAs with different levels of resource input to better understand the value added by increased resource input. Lack of expertise and available resources will limit capacity, but it is not known by how much. This is particularly relevant to small communities with limited budgets. Narrowing the focus of the VA to identify (or quantify) the tradeoffs or costs of doing nothing may be pragmatic. For example, California began its first state-wide adaptation strategy (see California Natural Resources Agency, 2009) while facing significant budget and resource challenges. Despite the staff's initial resistance to conducting a limited, qualitative vulnerability assessment due to limited time and staff resources, the adaptation planning process itself led staff to recognize the importance of doing VAs. This resulted in a recommendation to build the necessary research base and understanding among staff and in the broader community. The state-wide vulnerability and adaptation study currently underway in California (2010-2011) will require long-term investment in relevant research, education and training, as well as building of staff capacity. Combining public and private resources could be an important strategy for leveraging the resources to address climate change adaption needs.

Learning how to scale up VAs. More work needs to be done to better understand the challenges of scaling up and stitching together place-specific vulnerability studies for generalized results relevant to an entire sector or region. For example, how do we synthesize the impacts of heat waves on urban centers in general, while still paying attention to the social fabric and economic constraints of a specific area?

Defining system boundaries. Climate impacts can be localized, but the parties that need to be involved in responding to the impacts are often more widely distributed. The communities affected by climate change generally have little control or influence over the events creating these effects. For example, farmers in the Midwest can experience the effects of climate-induced drought, but adaptation may depend on development of drought-resistant seed varieties that international companies provide. Furthermore, the farmers' vulnerability might be increased by the inability to save seeds from season to season if their contract with a company does not allow seed-saving.

IV. Criteria for Evaluating Existing Vulnerability Assessments

Workshop participants developed a general set of criteria for evaluating VAs for use in the 2013

NCA, building from a strawman list presented at the workshop (see Table 1). The Table gives a list of many of the key components discussed in the following section, which summarizes criteria for structural, content, and communication components

Table 1. Modified Straw Man List of the Nature of Key Components and Criteria for Including VAs in the 2013 NCA		
Key Components	**Criteria for inclusion into the 2013 NCA**	
Focus of Study		
Topic area	Fits into one or more NCA topic areas.	Essential to include in a VA
Temporal scale	Linked to planning time of decision-maker at the relevant scale and sector; both near- and long-term planning timeframes.	Essential
Spatial Scale	Boundary clearly defined and at a resolution relevant to decision-making; or, in the case of a nested assessment, at the scale of the need for policy relevant evaluations.	Essential
Current Status of useful information available to conduct VAs		
Baseline data	Long-term data and trends (where available) and current status of economic, social, demographic, health, infrastructure and ecological information.	Essential
Existing stressors	Current state of key stressors relevant to the question to be assessed.	Essential
Adaptive capacity	Historical understanding about how the region or sector has adapted. Linkage to resource availability (e.g., equipment, financial, human).	Essential
Future Projections		
Scenarios	IPCC emission scenarios and local scenarios. Local scenarios should be linked/lined up to IPCC emission scenarios if possible. Consistent across relevant sectors (climate, land use, water, etc. as appropriate); need some scientific consensus.	Useful, but not essential
Models	Multiple models, state of the practice (see scenario criteria).	Useful, but not essential
Multiple stressors	Quantitative or qualitative analysis of how climate may interact with existing stressors.	Essential
Uncertainty	Clearly articulated, can be expressed quantitatively, qualitatively, or semi-quantitatively.	Essential
Vulnerabilities Evaluated (based on exposure, sensitivity, and adaptive capacity of each)		
Economic	Linked to social/cultural and/or ecological indicators through scenarios, models, or qualitative analysis where appropriate.	Depends on topic
Social and cultural	Linked to economic and/or ecological indicators through scenarios, models, or qualitative analysis where appropriate	Depends on topic
Ecological	Linked to social/cultural and/or economic indicators through scenarios, models, or qualitative analysis where appropriate	Depends on topic
Use in Decision-Making		
Community input	Key stakeholders, decision-makers, and vulnerable populations have input into the assessment process.	Essential
Linked to decision needs	Relevant to policy and management. Assessment designed to support specific decisions or to provide information that could be re-cast for various decision-making needs.	Essential
Outreach and communications	Key results and implications from the assessment are communicated to stakeholders and communities.	Depends on topic
Evaluation and adaptive learning	The assessment process is evaluated and is seen as an adaptive learning process.	Useful, but not essential

of a VA. Table 1 provides the criteria related to each component and includes if participants felt each criterion was essential to assess whether a VA was done effectively enough to be included in the 2013 NCA. For example, participants saw including the current state of key stressors relevant to the question to be assessed as essential for a VA to be included, whereas linking local scenarios to IPCC (Intergovernmental Panel on Climate Change) scenarios would be useful, but not an essential component to determine including a VA in the 2013 NCA.

a) Structural Components

Assessment unit. All assessments should clearly indicate whether they are focused on determining the vulnerability of a population, a geographic area, or a particular sector (or some combination of these factors). For inclusion in the NCA, assessments should fit into one of the NCA topic areas (sector, region, or cross-cutting topic) and this can serve as a primary screen for choosing whether or not to consider an assessment further.

Spatial scales. Spatial scales should be appropriate to the type of vulnerability. Spatial scales are important to consider, but criteria should be flexible enough to accommodate different needs across sectors. For example, issues related to water are often local, but energy issues can be regional or national. When addressing natural systems, spatial scales should be ecologically-relevant.

Temporal scales. As with spatial scales, the temporal scale for an assessment will depend on particular sectors. Temporal scales could be based on the planning time horizon that is used by the decision-makers for a particular sector. For example, water and natural resource planning might require a different time horizon than transportation.

A dynamic approach. Vulnerability assessments should consider the changing physical, social, and economic conditions in addition to projected changes in climate.

Scientific credibility. Scientific credibility should be distinguished from peer review, as the peer-review process does not determine the utility of information from a decision-making perspective. The gray literature is a great source for assessments that could be included in the NCA. When examining assessments that are not peer reviewed, it is important to determine the reliability and validity of the information source. A number of VAs from the gray literature can be highly valuable, as long as the NCA adopts a review process that maintains scientific credibility.

b) Content Components

Qualitative and quantitative information. Some dimensions of vulnerability can be quantified more easily for all sectors, such as exposure. Assessing adaptive capacity and sensitivity can require a more qualitative response. Quantitative information is not always available at the spatial scales at which decisions are made, so qualitative inferences may need to be made. Qualitative stakeholder input can help identify what people value, along with other information about adaptive capacity and sensitivity to impacts.

Baseline information. Effective use of community baseline information such as demographics, health status, environmental stressors, infrastructure, resource availability, and legal constraints to adaptation is necessary to determine current and future adaptive capacity, as VAs require an established baseline from which to measure change. Understanding the historical dimensions of vulnerability is also important, especially the causes behind current symptoms.

Box 1: Suggestions for Data Sets

- Community demographics (e.g., income, class, gender, age, language, literacy, property ownership)
- Public and community health
- Local history
- Income inequality
- Land tenure and use patterns
- Geography
- Social and linguistic isolation/integration
- Family structures
- Community support networks, organizations
- Natural resource based vs. other sources of income
- Cultural ideologies (e.g., concepts of what nature is, cultural cycles of time vs. linear time)
- Sustainability of community
- Communities of space and place
- Social memory and responses to extreme events
- Government structures
- Broader forms of governance
- Ecosystem variables (e.g., soil moisture, drought index)
- Infrastructure
- Population density

Detailed, integrated data sets.
Data used in VAs should be appropriate for the vulnerability and scale being addressed. Data sources should be clearly identified and any data to be shown or incorporated in the NCA should be available for public use (see Box 1 for examples of data sets and Figure 3 for a diagram that accounts for integrated data sets to measure vulnerability).

Ecological, social, and economic indicators and their associated vulnerabilities. Social, ecological, and economic factors can interact in important ways, and should be considered for all sectors but may vary in their degree of detail. Many sectors, such as agriculture and forestry, sit at the intersection of these three areas, and should make an effort to address these important interactions where possible. VAs can provide rich understandings of how natural-human systems are dynamically coupled; however, some sectors may require less consideration of coupled human-ecological systems (see Figure 4 for

a case study linking socioeconomic and ecological impact assessments).

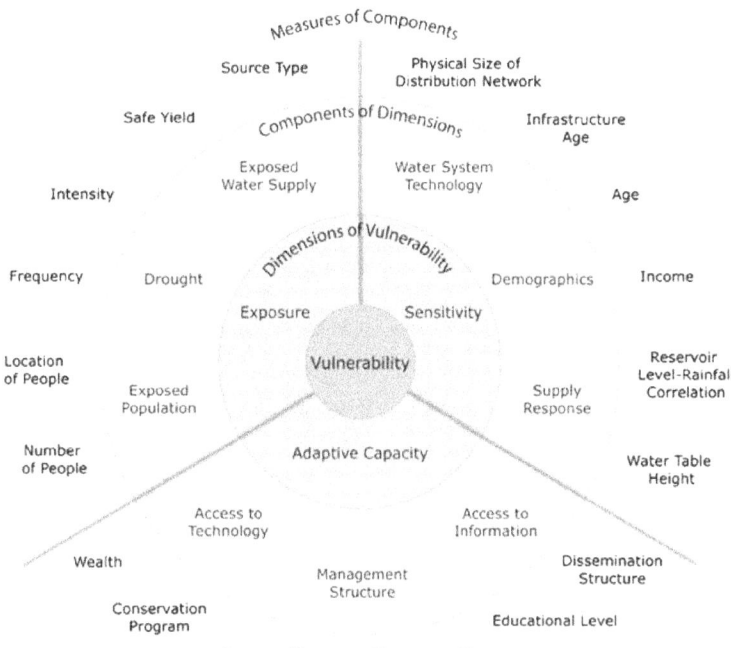

Figure 3. Hypothetical Vulnerability Scoping Diagram based on Human-Environment Regional Observatory Research Project, Example of four study sites in the U.S. exhibiting vulnerability of local water supply systems to the effects of drought. Source: Polsky *et al.* 2007:479.

Coastal Resilience Long Island: Adapting Natural and Human Communities to Sea Level Rise and Coastal Hazards

Coastal Resilience Long Island: Adapting Natural and Human Communities to Sea Level Rise and Coastal Hazards

Example of a methodology linking climate stressors (sea level rise) to hazards (storm surge) and integrating socio-economic and ecological impact assessments and spatial analysis tool.

Coastal Resilience Long Island: Adapting Natural and Human Communities to Sea Level Rise and Coastal Hazards

The Nature Conservancy (TNC) developed a demonstration project to help stakeholders on Long Island understand likely sea level rise (SLR) and coastal storm impacts and risks, visualizing them to identify management options that diminish losses and increase the resilience of natural and human coastal communities. A methodology and mapping tool were developed for analyzing the viability and adaptability of ecological resources potentially impacted by sea level rise and coastal storm impacts. Socioeconomic analyses were also applied to determine the potential consequences of SLR and storm surge hazards for human populations, providing information that allows managers to explore opportunities to minimize these consequences.

http://coastalresilience.org/

Figure 4. Coastal Resilience Long Island: Adapting Natural and Human Communities to Sea-Level Rise and Coastal Hazards.

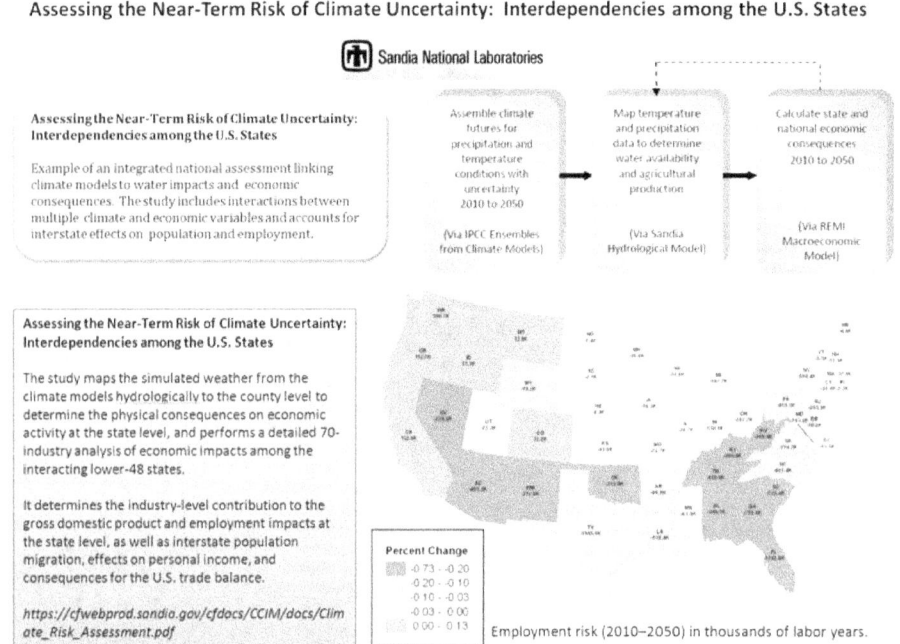

Figure 5. Assessing the Near-Term Risk of Climate Uncertainty: Interdependencies among the U.S. States

Uncertainty. Uncertainty can either be measured quantitatively or expressed qualitatively (see Figure 5 for a case study of an integrated national assessment measuring climate uncertainty). Quantitative measures such as likelihood and probability are often difficult to determine when conducting VAs of future climate change. For example, the U.S. spends about $100 billion/year on transportation services – how good and/or certain do researchers and decision-makers need to be about potential impacts and effects to guide those investments? Uncertainty can and should still be expressed in the absence of quantitative information by using qualitative statements from experts or by using multiple scenarios of possible futures. Defining level of confidence or uncertainty would enhance the VA conclusions.

Multiple stressors. Many impacts of climate may manifest themselves as interactions with other key stressors, which the 2013 NCA should begin to identify (see Figure 6 for an example of a study on shoreline change using multiple stressors). For example, increased drought from warmer, drier summers may increase wildfire risk in some forest systems. Additionally, migration of human populations from flooded coastal areas could place more demands on economic resources elsewhere. While it is not necessary to determine every possible stressor in every assessment, each assessment should address the key stressors that influence the sector or region that could potentially interact with

climate change. This also means including both the direct and indirect effects of climate change (such as effects on poor communities due to mitigation-driven energy policies or a rise in food prices) including market effects created by climate change that can also affect communities.

Climate projections. VAs that focus on climate as an important component should include some information about the projected changes in climate and its associated impacts and be consistent across the various components of the assessment. Where available or relevant, assessments should make use of the most up-to-date climate and associated impact models, such as being linked to or lined up with the IPCC projections. However, some excellent assessments might not include climate projections at all, but still look at vulnerability. Global climate models do not always provide the level of detail needed for regional or local planning. For regional assessments, regional models that are regionally created and validated can be more useful. Local assessments may have to rely on more qualitative information and expert opinion when models are not available on a relevant scale. When global climate models are used, however, a range of multiple models should be used to address uncertainty.

c) Communication Components

Mapping. Many existing VAs use mapping to describe current states and potential futures. Conveying information in this way can be useful when communicating with stakeholders. For example, a map of projected sea-level rise can help inform planning decisions for coastal areas. The 2013 NCA could improve upon these maps by bringing different map layers together (cumulative indices) and putting together series of maps that show vulnerability over time.

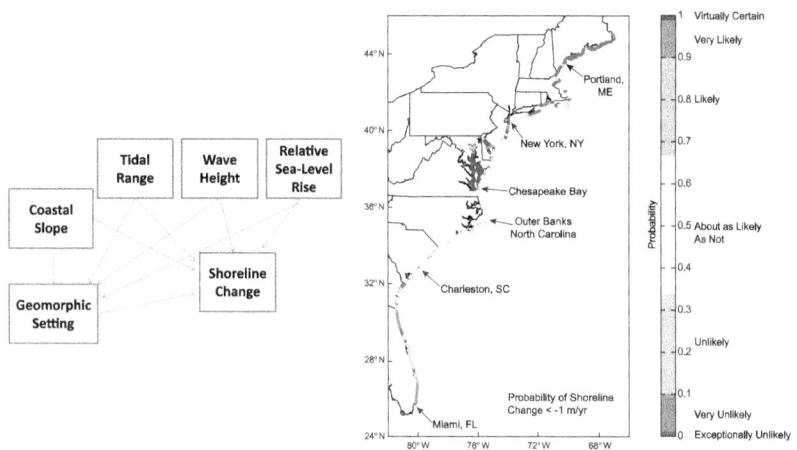

Figure 6. Left: Structure of the Bayesian network (BN) used to describe quantitatively the probability of different shoreline change rates given knowledge of several factors that define a particular shoreline setting. The BN approach allows users to evaluate the probability of a specific outcome based on causal relationships between a wider range of variables deemed important by users. Here the rate of relative sea-level rise, mean wave height, and tidal range are considered driving forces; the coastal slope and geomorphic setting are considered boundary conditions; and the shoreline change rate is considered to be the response variable or vulnerability indicator. Right: Map of the U.S. Atlantic coast showing the posterior probability of shoreline change < -1 m/yr using a BN approach. The probabilities are color-coded and labeled using IPCC likelihood terminology (Source: Gutierrez et al., 2011).

nity input. Researchers could also look at social networks to connect with communities; examples of networks to connect with include local watershed councils, cooperative weed management areas, planning groups, flood plain associations, co-ops, and green groups. Many current vulnerability assessments have actively involved stakeholders in regions and sectors. Lessons learned and best practices from these examples could be incorporated into the 2013 NCA process for engaging stakeholder groups (see Figure 7 for a case study using participatory research methods).

Decision-focused. Assessments can either take a top-down or bottom-up approach when linking to decision-making. Top-down approaches start with climate impacts and then identify what decision-making processes may be affected. Bottom-up approaches start with the decision-making process and identify potential impacts that may be relevant to that specific process (e.g., "client-based" approaches). Some assessments may be a combination of the two, using a top-down framework but gathering the majority of information from the bottom up. Regardless of the approach taken, assessments should be useful to end-users in making more informed decisions. In relation, researchers conducting VAs should be engaged with policymakers during the assessment process.

Stakeholder input. Local context and knowledge is essential for many decision-making processes and should be included as input into VAs. To ensure local knowledge and values are incorporated, the structuring and implementation of the VA process should include a bottom-up approach. Researchers doing VAs should be engaged with local communities during the assessment and data collection process. One way to do this is through conducting town-hall-meeting-style forums for local commu-

V. Vulnerability Assessment Framework for the 2013 NCA Report

a) Reframing Vulnerabilities for 2013

Workshop participants were charged with identifying key components of an integrated climate VA and providing input on what an integrated VA might look like in the regional and sectoral structure of the 2013 NCA Report. The group considered new approaches that would be efficient (built from existing assessments), dynamic (connected to updated NCA climate science, modeling, impacts, etc.), and most importantly, useful (relevant to current decision-making especially at local levels).

Leading with issues
Vulnerability information can be packaged and communicated in a manner that "leads" with the issues of concern regarding climate impacts on already-stressed or sensitive resources. This approach defines vulnerabilities based on known stresses. It can build on existing VAs for hazards and environmental threats, integrating climate change impacts and associated vulnerabilities into issue-focused, decision-relevant contexts.

Enhancing Gulf of Mexico Coastal Communities'
Resiliency through Participatory Community Engagement

Example of a qualitative study using participatory methods to
identify individual and community factors of social resilience.
Identifies potential indicators and techniques for use in
evaluating adaptive capacity in the NCA.

University of New Orleans

Social Resilience Assessment Methodology

➤ Traditional Knowledge
➤ Participatory Action Research
➤ Declaration of Principles
➤ Entree
➤ Semi- Structured Interviews
➤ Community Reception and Shrimp Boil
➤ Cross Coastal Resiliency Forum

Enhancing Gulf of Mexico Coastal Communities' Resiliency through Participatory Community Engagement

Using a Participatory Action Research (PAR) methodology, the University of New Orleans - CHART (Center for Hazards Assessment, Response and Technology) worked to identify and enhance community resilience in partnership with the residents of the Jean Lafitte area of Louisiana. The strategies that were used both documented and analyzed community member's knowledge of resilience, and indicators thereof, within the community following hurricanes Katrina and Rita (2005).

Some key findings for identifying and assessing adaptive capacity include:
• Indicator measurements of a community's vulnerability often miss community social systems of support and citizen engagement that has the ability to foster resilience in the face of high risks.
• The significance of strong place attachment and social networks is not adequately represented in our current understanding of resilience and adaptive capacity.

http://chart.uno.edu/projects/new-projects.aspx

Figure 7. Enhancing Gulf of Mexico Coastal Communities' Resiliency through Participatory Community Engagement

• Presenting vulnerabilities in the 2013 NCA from the perspective of the incremental and exacerbating effects of climate change on existing stressors would provide substantial benefit to the sectoral and regional reports from the use of existing robust and scientifically credible VAs developed for hazards and environmental stressors.

• Many existing assessments contain vulnerability information that is already understandable and familiar to decision-makers. This information could be used in the 2013 NCA to provide appropriate context to decision-makers and stakeholders for understanding, communicating and adapting to the potential effects of changing climate conditions.

Linking to decision processes
Participants emphasized repeatedly that the identification of vulnerabilities without obvious linkages to potential adaptation actions is not a constructive or useful framework for decision-makers (see Commentary 1 for connecting VAs to decision-makers). When issues of vulnerability are raised, there should be practical solutions to match the vulnerabilities. Additionally, providing decision-makers with a basic understanding of vulnerability would provide multiple benefits, such as prioritizing research

agendas or enhancing understanding of population dynamics.

• If vulnerabilities are presented in the 2013 NCA from the perspective of the exacerbating effects of climate change on existing stressors, the planning and policy structure currently in place to address those stressors can serve as an initial decision framework for linking vulnerabilities to adaptation actions.

• Participants stressed the need to incorporate vulnerability information in actionable contexts, using positive language so people are willing to take action to work towards solutions, instead of creating a doom-and-gloom scenario (see Figure 8 for a case study linking VAs with adaptation plans).

• Too many decision-makers view climate adaptation as an additional cost when planning and when decisions about new policy and investments are being made. Decision-makers generally do not make tangible connections between climate adaptation as a strategy and their current decision-making processes. Framing vulnerability in specific decision-relevant contexts provides those connections and provides

Commentary 1

Susi Moser, Perspective on Applied Vulnerability Assessments at the Local Level

At the local level, there is a recent increase in interest to develop adaptation plans and strategies. In cooperation with the Local Government Commission and the Geos Institute, two California Counties (San Luis Obispo and Fresno counties, including various municipalities in each) were approached in 2010 to initiate a local discussion about adaptation through a stakeholder-intensive, participatory process (for more information, see Moser and Ekstrom, 2011). The social system component of the project (community, economy, infrastructure and social services) was supported by background research on vulnerabilities of each county (for more information, see San Luis Obispo, http://www. lgc.org/adaptation/slo/ and Fresno, http://www.lgc.org/adaptation/slo). Among the most important insights from the experiences in these two economically and politically very different locations are the following:

Vulnerability assessments – as direct and locally experienced and verifiable mirrors of a community – are very useful conversation starters, in many ways more so than climate change science and projections. A structured, facilitated dialog around vulnerabilities lends itself for stakeholder engagement and for integration of local knowledge into a scientific VA ultimately resulting in an improved understanding of local threats and assets. To the extent that process can be done at a time that allows for integration with other planning processes, there is a real chance to integrate vulnerability thinking into policy-making. However, vulnerability assessments don't always have a clear decision-maker audience as existing planning processes don't necessarily all include such thinking and expertise (exceptions may be public health and emergency management). Thus, the outcomes of a VA are not easily integrated into decision-making – for example, in priority-setting. As VAs inherently surface ethical aspects and deep-seated societal challenges and discontents, some audiences will want to hear and look at them, while others would rather avoid them. Pointing to assets and solutions to improve local situations regardless of climate change is thus critical.

actionable information that can support their job performance.

- By demonstrating the importance of specific elements in the natural-human system, VAs make it clear what kind of data decision-makers need. For example, highway department managers may learn how underground utilities are vulnerable to flooding and reassign crew priorities

during flood events to prevent disruption of internet, telephone, and electricity. Knowing the elevation of places where the utilities exit the ground would be useful data to have.

Addressing adaptive capacity

As a major component of VAs, adaptive capacity will need to be addressed in some coherent and consistent manner throughout the 2013 NCA. At

Miami-Dade County Roadmap for Adapting to Coastal Risk

Miami-Dade County Roadmap for Adapting to Coastal Risk

Example of an issue-driven methodology assessing vulnerabilities that are relevant to specific decision-making processes. Uses a participatory process, relying on local expertise and knowledge to supplement existing risk and vulnerability information on hazards and climate impacts.

Miami-Dade County Roadmap for Adapting to Coastal Risk

Miami-Dade County mainstreamed hazards and climate information into existing planning and operations (land use, infrastructure, public safety, *etc.*). In partnership with NOAA's Coastal Services Center, the Miami-Dade County Office of Sustainability developed and implemented a participatory vulnerability assessment process, "Roadmap for Adapting Coastal Risk" with a goal of identifying efficient and effective mechanisms to integrate issues into existing decision processes.

The Roadmap effort was not designed to create a new or stand-alone assessment product or plan but to engage County staff and stakeholders in hands-on activities that put a hazards and climate lens on issues of relevance to them such as water availability, stormwater management, social services, natural resource conservation, capital improvements, emergency management and public transportation.

http://www.csc.noaa.gov/digitalcoast/training/roadmap/index.html

Roadmap Process

1) **Getting Started**—define community goals and objectives and highlight priority issues and drivers for consideration throughout the assessment.

2) **Hazards Profile**—explore relevant hazards, climate trends and potential impacts as a starting point for considering community vulnerabilities.

3) **Societal Profile**—evaluate strengths and vulnerabilities of local population through analysis of demographics combined with local knowledge and expertise.

4) **Infrastructure Profile**—identify the strengths and vulnerabilities of the built environment through geographic analysis combined with local knowledge.

5) **Ecosystem Profile**—consider the strengths and vulnerabilities of important natural resources through data assessment combined with local expertise.

6) **Taking Action**—explore opportunities and challenges for risk-reduction through education, planning and regulatory processes.

Figure 8. Miami-Dade County Roadmap for Adapting to Coastal Risk

a minimum, adaptive capacity needs to be defined (using a common starting definition) for the NCA sectors and cross-cutting issues below. If at all possible through the use of existing VAs and expert knowledge, potential "indicators" of adaptive capacity should be identified and used to help frame the vulnerability assessment narrative in the NCA report.

b) Developing an Integrated and Iterative Approach to Vulnerability Assessment in the 2013 NCA

Workshop participants identified the need to take an integrated and iterative approach to VAs in the NCA that includes common linkages across sectors and geographies, considers vulnerability from a holistic "systems" perspective wherever possible, and is adaptive as more successful VAs are completed (see Figure 9 for an example of an integrated assessment). Below are some key considerations for this approach.

Linking to common scenarios

Recognizing that the upcoming NCA will depend heavily on existing studies and resources, VA parameters and methods will not be fully consistent across sectors, systems, and regions. Including scenarios of possible futures that take other factors besides climate change into account could help with integration within sectors and regions and across the entire assessment. Considerations include

- Consistent scenarios for climate, population, and other future changes will need to be established and agreed upon early in the process. This should be coordinated with other NCA working groups early in the process.
- Existing vulnerability studies may not include future conditions, and if they do, are likely to be based on different scenarios. These studies can still be used and referenced in the NCA, although expert judgment should be applied to provide a qualitative discussion linking them to the future conditions represented in the consistent scenarios.

Building on existing climate impact assessments
- Climate impact assessments should play a critical role in the VA framework. Impact assessments are generally designed to identify the potential effects of changing climate conditions on human populations and natural systems, providing a critical foundation for identifying vulnerabilities. Relevant components of many

Integrated Assessment for Effective Climate Change Adaptation Strategies in New York State

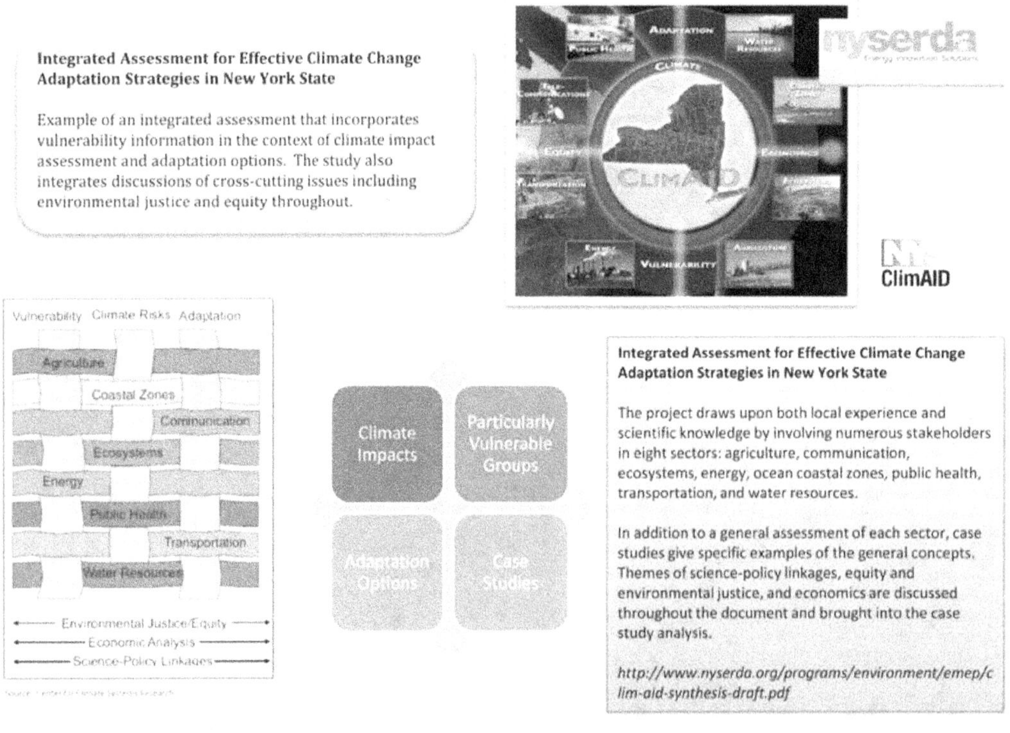

Integrated Assessment for Effective Climate Change Adaptation Strategies in New York State
Example of an integrated assessment that incorporates vulnerability information in the context of climate impact assessment and adaptation options. The study also integrates discussions of cross-cutting issues including environmental justice and equity throughout.
nyserda
ClimAID
Vulnerability Climate Risks Adaptation
Agriculture
Coastal Zones
Communication
Ecosystems
Energy
Public Health
Transportation
Water Resources
Environmental Justice/Equity
Economic Analysis
Science-Policy Linkages
Climate Impacts
Particularly Vulnerable Groups
Adaptation Options
Case Studies
Integrated Assessment for Effective Climate Change Adaptation Strategies in New York State
The project draws upon both local experience and scientific knowledge by involving numerous stakeholders in eight sectors: agriculture, communication, ecosystems, energy, ocean coastal zones, public health, transportation, and water resources.
In addition to a general assessment of each sector, case studies give specific examples of the general concepts. Themes of science-policy linkages, equity and environmental justice, and economics are discussed throughout the document and brought into the case study analysis.
http://www.nyserda.org/programs/environment/emep/clim-aid-synthesis-draft.pdf

Figure 9. Integrated Assessment for Effective Climate Change Adaptation Strategies in New York State impact assessments include

- Identification of multiple effects or stressors associated with future climate conditions.
- Estimated exposure of key systems and resources to the projected effects.
- Modeled future system/resource conditions (populations, natural systems, and resources).
- Modeled sensitivity of key resources to projected effects.

A critical first step is to conduct a full review in collaboration with the regional and sectoral working groups to determine what climate impact assessments are available and most relevant for use in VAs.

Progressing toward a "systems" approach
Vulnerability assessments need to be iterative and flexible, building a framework that links together different studies. The use of existing studies and resources in the 2013 NCA will limit the ability to fully implement an integrated "systems" approach to VA. Alternatively, some key considerations identified by the working group can provide a starting point for identifying and evaluating existing assessments or case studies that demonstrate a more holistic approach to vulnerability assessment. Assessments need to go beyond the physical and biological aspects of exposure and sensitivity to include the social, economic, institutional, and regulatory aspects of vulnerability. It will be important to ensure that the 2013 NCA includes representation of some of the critical interconnections between social and natural system vulnerabilities across different sectors and scales.

Generalizing site-specific assessments - Aggregating site-specific VAs could contribute to broader vulnerability assessment information for use at regional and national scales. Using results from fully integrated local studies, broader regional stories could be developed around different types of human-environment interactions for the 2013 NCA. This approach would be particularly appropriate for existing VAs in public health and natural resources where the interactions between multiple stressors are site-specific and highly complex.

Human-environment dependencies – Considerations of human-environment interactions and dependencies are critical to evaluating vulnerability. For example, populations dependent on local resources such as fisheries or agriculture are particularly vulnerable to perturbations from hazards and environmental stressors. In addition to livelihood

impacts associated with threats to the environment, many of these populations also face related issues affecting their adaptive capacity (health, cultural identity, social networks). The vulnerability of human populations can be closely linked to the ecosystem "services" provided by natural resources (see Figure 2 for example of human-environment interactions and vulnerability to climate change impacts).

Environmental justice and marginalized populations – Definitions of components in a system-based approach must start from the knowledge that natural disasters, even slow-onset disasters, affect the poor and the powerless the most. Vulnerability is derived from poverty, lack of resources or lack of access to adapt and hence, compromised adaptive capacity. A system-based approach relies on the best possible description of the socioeconomic and cultural context of the communities, neighborhoods, and families in the units of description. Case studies or examples of integrated assessments that include analysis of societal values and institutional, political, economic, and power dynamics could provide critical information for evaluating the human side of adaptive capacity. Social science studies addressing these factors (especially those grounded in local knowledge) could be integrated to help bridge this adaptive capacity gap in the current vulnerability assessment literature.

Societal responses – Ideally, VAs for future climate assessments will also factor societal responses into the analysis. Given the time and resource limitations of the 2013 NCA, it is unlikely that a comprehensive evaluation of societal responses can be included although it could be feasible to include relevant studies that could demonstrate the effects of specific societal responses on future vulnerabilities. Examples might include analysis of economic benefits or costs avoided through policy changes or projected effects of an institutional response such as the transfer of economic risk from one group to another

c) Vulnerability Assessment Process for the 2013 NCA

Some of the key activities that need to be considered in moving forward with VAs for the 2013 NCA are highlighted below.

Develop lexicon for vulnerability assessments in the NCA - Recognizing that there are numerous definitions and distinctions across disciplines, develop a lexicon for use in the NCA process. The document will provide reference definitions for the language to be used in the NCA and will serve as a resource for communicating across disciplines in the development and integration of VA information.

Develop hazards/stressors taxonomy – A taxonomy should identify and categorize hazards, threats, and environmental stressors in a manner that could support the systematic evaluation of current risks, anticipated climate change impacts, and potential effects of societal changes. To be used in conjunction with the lexicon, the taxonomy provides for use of common "hazards" language in the NCA and serves as a key resource for organizing, evaluating, and communicating VA information.

Develop decision-drivers matrix – Vulnerability assessment information used in the 2013 NCA should be developed to inform decision-making processes at national, regional, and local levels. A matrix listing targeted "decision drivers" should be developed to help guide the VA process toward this goal. With direct stakeholder input, identify at least one specific decision-driver at each scale (national, regional, and local) for every sector and integrated cross-boundary topic in the NCA. Illustrative examples of decision drivers might include

- Department of Defense establishing climate adaptation priorities for military installations (national)
- Federal Highway Administration updating highway planning and design guides (national)
- State of Maryland establishing coastal conservation and restoration priorities for marsh migration (regional)
- City of Milwaukee updating long-range stormwater management planning and design standards (local)

- The American Red Cross developing long-range disaster response and shelter strategies (national)
- The decision-driver matrix would not represent all uses for vulnerability information in the NCA, but would provide essential issue-based framing for evaluating the applicability of existing assessment resources in the 2013 NCA.

Expanded scope of literature reviews - The NCA VAs need to build on the strengths of previous approaches, especially those assessments that cross the divides of sectors, disciplines and geography. Because VAs operate at the intersection of science and policy, it is critical to cross disciplinary boundaries, especially as social science is under-represented in existing assessments. Some possibilities include policy and adaptive management literature; anthropology literature on social systems, human dimensions of change, and human-environment coupled systems; political economy literature on integrated assessments; and hazards literature on risk amplification. Additionally, the gray literature contains many VAs that could be very valuable; the NCA would need to adopt a review process to maintain the scientific credibility to include these studies.

Develop an action-focused communication strategy – Vulnerability assessment information should be framed positively to support adaptation actions in the 2013 NCA. An overall communication strategy should be developed for presenting vulnerability information in a constructive and consistent manner throughout the report. Comparative studies, qualitative assessments, and adaptive capacity evaluations represent different types of communication challenges to be addressed in the report. For example, assessments of adaptive capacity related to marginalized populations should be framed in

Commentary 2

Robin Bronen, University of Alaska, Fairbanks

Newtok is a traditional Yup'ik Eskimo village located close to the Bering Sea in western Alaska. The village's ancestors have lived on the Bering Sea coast for at least 2,000 years. Approximately 350 people currently reside in the community. The combination of decreased arctic sea ice, thawing permafrost, and increased extreme weather events has created a humanitarian crisis in Newtok. The Newtok Traditional Council has been documenting these ecological changes since 1983. The community has documented the loss of critical basic necessities and infrastructure due to accelerated rates of erosion, including the barge landing facility necessary for the delivery of fuel to power the electricity in the community, the loss of the village dump site, and the loss of potable water due to salination. In 2004, the community voted to relocate because the traditional methods of erosion control and flood relief could no longer protect the community (for further information, see Bronen 2011).

terms of conditions that enable or constrain actions, as opposed to focusing on population characteristics.

Create compelling stories. Vulnerability assessments need to include case studies grounded in impacts at the local scale. Local-level case studies and VAs can provide narratives of compelling stories to create connections between different sectors and regions. The NCA needs to include concrete examples of what communities have already done to identify and address major vulnerabilities faced, providing narratives of compelling stories that influence people's opinions. Compelling narratives could make the NCA much more useful to people and much more likely to be used (see Commentary 2 for an example).

VI. A Sustainable Process for Future NCA Vulnerability Assessments

While it may be unrealistic to develop new VAs for 2013 or to fully integrate existing disparate VAs into a cohesive framework for 2013, one of the goals of the workshop was to identify ideal core elements of an integrated assessment that could serve as an implementation framework for future NCAs.

Participants envisioned the VA component of the NCA as an evolving process so that each NCA would learn from the previous NCA, particularly about the Nation's vulnerability to climate change. The NCA would provide guidance on the methods of conducting regional and local VAs, especially to ensure that they are cross-sectoral and include multiple stresses, but also how those assessments would provide information on the national vulnerabilities. This latter aspect might include identifying higher-order questions that link on-the-ground issues with the adaptive capacity of federal institutions. Vulnerability assessments would take the longer-term view, even though the reporting period for the NCA is every four years, and could periodically focus on issue or theme-based studies to facilitate scaling-up the vulnerabilities. Primarily, the participants suggested that the sustainable process for future NCA vulnerability assessments could be a civic discovery process where engaged stakeholders facilitated the broader learning and understanding about vulnerability in communities across the U.S.

The NCA as a roadmap - Participants emphasized the potential for a sustained NCA process to provide

a national framework that can be used to inform and guide regional and local efforts in the development of VAs (see Figure 10 that demonstrates the flow amongst the NCA local VAs). The NCA could provide a template that could be used across communities for designing and implementing local scale VAs. It would address the following:

- Looking at the VA process as a continuum providing guidance on approaches for assessing vulnerabilities to current and future impacts.
- Include community engagement strategies that encourage participants to develop, implement, and evaluate solutions.
- Provide an end-to-end approach that links climate impacts-vulnerability-adaptation-mitigation.
- Include resources such as data, mapping, literature review, and basic methodology that can be layered on local efforts.

Issue- or theme-based studies – A series of issue- or theme-based VAs could be developed for the NCA that are fully integrated (e.g., water, health). Different theme-based assessments could be rolled out both nationally and regionally, with different years highlighting the needs of different themes.

Longer-term perspectives – Integrated VAs for the NCA should look beyond ten years, as most vulnerability assessments do. Temporal issues to consider include past and projected long-term future changes in data availability, analytical capacity, population, technology, economics, institutions, behavior and culture, which are not static.

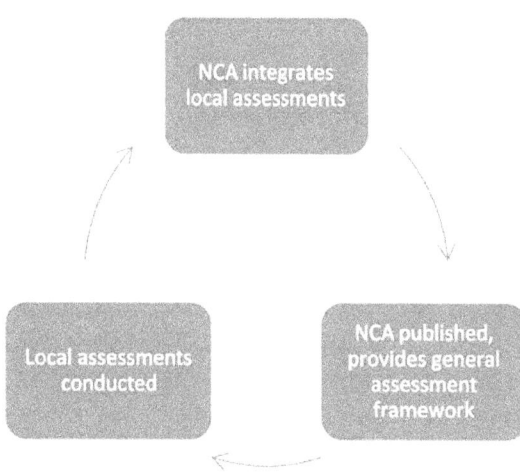

Figure 10. Sustained process using NCA as a roadmap.

Identification of cross-scale thresholds or tipping points – Integrated VAs for the NCA should address thresholds and tipping points at different scales, from local to national. Connections need to be made between local and surrounding vulnerabilities and how those vulnerabilities translate to areas farther away. For example, an integrated VA could assess how economic risk and vulnerability are transferred across scales.

Expand and leverage assessments – Increasing the number of VAs, the people working on them (including social scientists, city planners and leaders, private sector representatives) and the number of people providing input will help decentralize the process. A more distributed process that involves stakeholders at all levels, including local leaders, local government, private sector, *etc.* would be effective. It would be useful to provide a framework that builds capacity for future NCA efforts.

Highlight and integrate higher-order questions. One possible approach for addressing vulnerability in the NCA is to have a set of key system-wide questions that would be addressed over a four-year period as a driver for research priorities. In the natural resources sector, questions could focus on an institution or system rather than an individual unit, such as:

* What is the vulnerability of our federal public lands, *e.g.*, the National Park System as an institution that responds and adapts to climate change?
* What is the vulnerability of our non-federal lands, *e.g.*, what happens to conservation easements as private land managed for multiple ecosystem services under a changing climate?

The NCA could use these questions to consider higher-order effects, both external effects and resulting impacts.

A civic discovery process. Assessing vulnerability in the NCA should be an ongoing adaptation and learning process that involves many groups in addition to the science community. This involvement would include extensive consultation with diverse stakeholders to keep the context relevant. The NCA should partner with local researchers and universities and engage with local officials and organizations.

VII. Workshop Perspectives

Three workshop participants provided perspectives on the discussions at the end of each day of the workshop (see Agenda for list of speakers, Appendix A). Participants' summarized key input received throughout the workshop, such as the need for clear definitions and terminology at every stage of an assessment, including "risk" and "impact", and the need to address the legal, policy and institutional barriers that constrain adaptation actions.

Related to these barriers is also the need to understand the constraints on adaptation options. Social "willingness" to undertake or support the actions needed to respond to climate change is influenced by individual and community values, which can vary widely across individuals, geographic locations, and cultures. Action to address climate change must consider social, economic, biological, and physical constraints. Climate change effects cannot be assessed in isolation; the NCA and decision-makers have to consider the entire complex of stressors, not develop action items specifically for climate stressors.

The participants also reflected on the theme of communication being critical to encouraging stakeholders' engagement in the NCA and that climate is not necessarily the topic to lead with to engage people's interest. It might be better to lead with impacts of climate change that are of interest, *e.g.*, water in the Southwest. VAs provide an excellent tool for engagement to lead with what is vulnerable from climate change and the possible futures, rather than start with the climate science. The process of conducting a VA is an approach to help discuss potential impacts and adaptation options. VAs can serve a role in broader conversations about climate change by focusing on what is going on where people live and the environment they are familiar with, which may already be experiencing impacts due to climate change.

Education is essential in providing information about vulnerability and providing tools that can be used to assess options. The goal of putting long-term processes into place to provide ongoing exchange of information and reaching out to a broader community will be critical to the success of the NCA. This requires encouraging a diversity of views and opinions. The NCA in general offers a huge opportunity to build a national conversation about adaptation and mitigation. The 2013

NCA can be used as a teachable moment. A civil society conversation needs to engage groups such as Chambers of Commerce, industry (e.g., energy), local stakeholders, nongovernmental organizations (NGOs), opinion leaders in local communities. The goal is to build a community of stakeholders.

In particular, the NCA should reach out to tribes who are greatly motivated to be involved in the NCA and associated VAs. Tribes are faced with significant challenges in addressing climate change. Tribal lands cannot be "moved" to adapt to the changes. It is important to draw on the traditional ecological knowledge to address management options.

The reflections also echoed the need to reach decision-makers at all levels. While VAs can be conducted at various scales, most adaptation actions will occur at the local level. Local governments, tribes, private businesses, (NGOs), and private individuals all play critical roles in responding to climate change. A challenge for the NCA is determining the frame of reference and scale for providing information. Will the NCA analyses focus on providing information that has decision and/or policy relevance to local-, regional-, and national-scale decision-makers? Can the NCA provide a framework for assessing vulnerability or developing adaptation options? What is the balance between what is feasible to provide in a national assessment and information that can be used to guide local actions?

Participants suggested that the NCA could provide guidance on the core elements of VAs, particularly in providing context for conducting analyses at these much finer scales. Assistance in data aggregation issues would also be useful. What data can be fed both "up and down" to incorporate into vulnerability assessments? While local governments and other entities collect data at local scales, locals also rely on disaggregated data from state and federal sources to assist in these analyses. If the NCA is going to provide guidance for VAs, it needs to be dynamic, flexible and cross-sectoral.

Participants also reminded the workshop group and the NCA team that the potential role of VAs for the 2013 NCA is likely to be much more limited than for the sustainable process. Distinguishing what is feasible for the 2013 NCA versus the long-term process is important for planning work and resource allocation. A patchwork of VAs has been done for

a variety of objectives. Collectively they provide a good overview of the concerns of local communities, since most have been done at the local scale. For the 2013 NCA, reflecting on the lessons from that collective knowledge may be the best we can do. Depending on resources, it might be useful to identify where regional scale VAs would be most valuable for 2013. In the longer term, it might be important to take a broader look across the Nation. Do we need VAs conducted at the national scale or should we focus on regional and/or sectoral scale assessments that can be aggregated?

Finally, participants provided ideas on prioritizing the need for VAs for the NCA. Two possible options to consider are to 1) focus on the most vulnerable communities; and 2) focus where major infrastructure choices need to made that have long-term implications (i.e., they are built to last 40-50 years), such as coastal areas. Choose topics where ignoring climate change effects could result in maladaptive decisions.

VIII. Workshop Conclusions

Developing a **system-based approach** that connects regions, sectors, and multiple stressors in a more cohesive picture of the effects of climate change on human and natural systems will help the 2013 process of stitching together the extant information and facilitate the evolution of the sustaining NCA process of VAs. The systems approach would benefit from coordination across the working groups to identify an agreed upon lexicon, including definitions, as well as the suite of climate-economic-social scenarios and how uncertainty will be described.

The importance of **integrated assessments** was also clear. Understanding the interactions between stressors is critical, as is looking across sectors. Vulnerability assessments need to look more broadly than a particular sector or topic. The emphasis on cross-cutting themes in the planning for the NCA reflects this increased emphasis on integrated analysis, while realizing the challenges of conducting such analyses.

For the 2013 NCA, preliminary work to develop **links to decision-making** would ensure the successful application of information from the VA at the regional and local levels. Local context and knowledge is essential for many decision-making processes and should be included as input into the NCA vulnerability assessment. The 2013 NCA should include both an **action-focused communication** strategy that encourages stakeholders' engagement in the NCA and helps to build a community of stakeholders and a bottom-up approach to incorporate community input and participation, providing communities with a sense of empowerment and ownership of the NCA process.

The importance of equity, environmental justice, and institutions are also dominant in any discussion about vulnerability. Understanding the potential effects of climate change on people and natural systems is essential to making fair and informed decisions about adaptation options. It was emphasized throughout the workshop that often the most vulnerable human populations are difficult to identify and usually have little political power. Therefore, a fine-toothed knowledge of the social structure, the history, and the demographic composition and its "patchiness" of the population vulnerable to climate change impacts is essential to depict in a sound VA.

Assessing vulnerability in the NCA could be an **ongoing adaptation and learning process** that engages many sectors, regions, and institutions in discussions about vulnerability to climate change and potential adaptation options. Such a process could operate as a continuum providing guidance on approaches for assessing vulnerability to current and future impacts, providing guidance on the qualitative and quantitative methods to assess that vulnerability and a process to synthesize the results into guidance about the risks and options for mitigation and adaptation.

Acknowledgments

A special thanks to the U.S. Department of Agriculture Forest Service and the Centers for Disease Control and Prevention for co-sponsoring the NCA Vulnerability Assessment Workshop and to the U.S. Environmental Protection Agency for sponsoring the Workshop White Paper.

Works Cited:

Adger, W.N. 2006. "Vulnerability." *Global Environmental Change* 16:268–281.

Balbus, J.M. and C. Malina. 2009. "Identifying Vulnerable Subpopulations for Climate Change Health Effects in the United States." *Journal of Occupational & Environmental Medicine* 51(1):33–37.

Bronen, Robin. 2011. "Climate-induced Community Relocations: Creating an Adaptive Governance Framework Based in Human Rights Doctrine." *New York University Review of Law and Social Change* 35(2).

California Natural Resources Agency. 2009. "The California Climate Adaptation Strategy 2009: A Report to the Governor of the State of California." Sacramento, CA: Natural Resources Agency.

Center for Hazards Assessment, Response and Technology. On-going project. "Enhancing Gulf of Mexico Coastal Communities' Resiliency through Participatory Community Engagement." University of New Orleans, LA. Website, http://chart.uno.edu/projects/new-projects.aspx.

Climate Change Public Health Impacts Assessment and Response Collaborative. 2007. "Public Health Impacts of Climate Change in California: Community Vulnerability Assessments and Adaptation Strategies." Sacramento, CA: California Department of Public Health. Electronic document, http://www.ehib.org/papers/Heat_Vulnerability_2007.pdf.

Downing, T.E. and A. Patwardhan. 2004. "Assessing Vulnerability for Climate Adaptation." In *Adaptation Policy Frameworks for Climate Change: Developing Strategies, Policies and Measures,* B. Lim and E. Spanger-Siegfried (eds.), Cambridge University Press, Cambridge, UK, pp. 67–89.

Gutierrez, Benjamin, Nathaniel Plant and E. Robert Thieler. 2011. "A Bayesian Network to Predict Coastal Vulnerability to Sea-Level Rise." *Journal of Geophysical Research* 116(2).

Local Government Commission. On-going project. "Fresno County Climate Change Adaptation." Fresno, CA. Website, http://www.lgc.org/adaptation/fresno/.

Local Government Commission. On-going project. "San Luis Obispo County Climate Change Adaptation." San Luis Obispo, CA. Website, http://www.lgc.org/adaptation/slo/.

Mills, David, Cameron Wobus and Kristie Ebi. 2011. "Vulnerability White Paper: In Support of the National Climate Assessment's Vulnerability Assessment Workshop." Prepared by Stratus Consulting Inc. and ClimAdapt, LLC on behalf of the U.S. Environmental Protection Agency.

Moser, S.C. and J.A. Ekstrom. 2011. "Taking Ownership of Climate Change: Participatory Adaptation Planning in Two Local Case Studies from California." *Journal of Environmental Science and Studies* 1(1), in press.

National Oceanic and Atmospheric Administration's Coastal Services Center. On-going project. "Miami-Dade County Roadmap for Adapting to Coastal Risk." Charleston, SC. Website, http://www.csc.noaa.gov/digitalcoast/training/roadmap/index.html.

The Nature Conservancy. On-going project. "Coastal Resilience Long Island: Adapting Natural and Human Communities to Sea Level Rise and Coastal Hazards." Arlington, VA. Electronic document, *http://coastalresilience.org*

New York State Energy Research and Development Authority. 2010. "Integrated Assessment for Effective Climate Change Adaptation Strategies in New York State." New York. Electronic document, http://www.nyserda.org/programs/environment/emep/clim-aid-synthesis-draft.pdf.

Polsky, Colin, Rob Neff and Brent Yarnal. 2007. "Building Comparable Global Change Vulnerability Assessments: The Vulnerability Scoping Diagram." *Global Environmental Change* 17:472–485.

Sandia National Laboratories. 2010. "Assessing the Near-Term Risk of Climate Uncertainty: Interdependencies among the U.S. States." New Mexico and California. Electronic document, https://cfwebprod.sandia.gov/cfdocs/CCIM/docs/Climate_Risk_Assessment.pdf.

Appendix A: Agenda

DAY 1

8:00	**Registration**
8:30	**Welcome and Overview** George Luber, Centers for Disease Control and Prevention (CDC) Lander Stoddard, CDC
8:40	**Overview of the National Climate Assessment** Kathy Jacobs, Office of Science Technology and Policy (OSTP)
9:25	**Workshop Objectives** George Luber, CDC
9:30	**Quick Break**
9:40	**Panel on Vulnerability Assessments: Setting the Stage** Sandy Eslinger, NOAA, Moderator Margaret Davidson, NOAA Susanna Hoffman, Hoffman Consulting Linda Joyce, US Forest Service Jonathan Patz, University of Wisconsin-Madison
10:40	**BREAK (refreshments provided)**
10:55	**Break-out Session 1: Criteria for Vulnerability Assessments** *Objective: Identify key components of a vulnerability assessment and criteria on which to evaluate the applicability of existing vulnerability assessments to sectoral and regional analyses in the 2013 NCA.*
12:10	**Lunch (on your own)**
1:10	**Plenary Report-Out on Break-out Session 1**
2:10	**Break-Out Session 2: Relevance of Current Vulnerability Assessments** *Objective: Using the key components and criteria developed in Breakout Session 1, evaluate the "current state" of vulnerability assessments for informing sectoral and regional analyses in the 2013 NCA, highlighting relevant existing assessments and identifying significant gaps.*
3:25	**BREAK (refreshments provided)**
3:40	**Plenary Report-Out on Break-out Session 2**
4:40	**Perspectives on Day 1** Daniella Hirschfeld, ICLEI – Local Governments for Sustainability Mike Savonis, Department of Transportation Kathleen Sloan, Yurok Tribe
5:00	**Wrap-Up**

DAY 2

8:30 **Recap of Day One and Agenda Preview**
George Luber, CDC

9:00 **Break-out Session 3: Incorporating VAs into the NCA**
Objective: Identify key components and guiding principles for integrating vulnerability assessment across sectors and regions in the NCA to help quantify and prioritize risks on a consistent national scale.

10:30 **BREAK (refreshments provided)**

10:45 **Plenary Report-Out from Break-Out Session 3**

11:45 **Lunch (on your own)**

12:45 **Applied Vulnerability Assessments: Lessons Learned**
Julie Maldonado, USGCRP, moderator
Robin Bronen, Alaska Immigration Justice Project
Susanne Moser, Moser Associates
Bruce Stein, National Wildlife Federation

1:45 **Break-out Session 4: A sustained NCA**
Objective: Design an approach to vulnerability assessments for the NCA sustained process.

3:00 **BREAK**

3:15 **Plenary Report-Out from Break-Out Session 4**

4:15 **Perspectives on Day 2**
John Hall, Department of Defense
Kirstin Dow, University of South Carolina
Peter Frumhoff, Union of Concerned Scientists

4:45 **Next Steps and Closing Comments**
Linda Langner, US Forest Service
Kathy Jacobs, OSTP

Appendix B: Participant List

George Backus, Sandia National Laboratories_
Jonathan Balbus, National Institute of Environmental Health Sciences
Doug Bausch, Department of Homeland Security
Charles Beard, Center for Disease Control and Prevention
Leslie Brandt, Northern Institute of Applied Climate Science, USDA Forest Service_
Robin Bronen, Alaska Immigration Justice Project / University of Alaska, Fairbanks
Lynne Carter, Louisiana State University_
Dave Cleaves, US Forest Service
Emily Cloyd, National Climate Assessment
Camille Coley, Florida Atlantic University
Marge Davenport, US Geological Survey_
Margaret Davidson, National Oceanic and Atmospheric Administration
Kirstin Dow, University of South Carolina / Carolinas RISA
Cathy Dowd, USDA Forest Service_
Hallie Eakin, Arizona State University
Paul English, California Department of Public Health
Sandy Eslinger, National Oceanic and Atmospheric Administration
Daniel Ferguson, Climate Assessment of the Southwest (CLIMAS) / University of Arizona
Shirley Fiske, University of Maryland
Peter Frumhoff, Union of Concerned Scientists
Bryce Golden-Chen, National Climate Assessment
Nancy Green, Fish and Wildlife Service
John Gross, National Park Service
John Hall, Resource Conservation and Climate Change, Department of Defense
Larry Hendrix, National Tribal Environmental Council
Daniella Hirschfeld, ICLEI – Local Governments for Sustainability
Jennifer Hoffman, EcoAdapt
Susanna Hoffman, Hoffman Consulting_
Kathy Jacobs, National Climate Assessment / White House Office of Science and Technology Policy
Bruce Jones, US Geological Survey
Linda Joyce, US Forest Service, Rocky Mountain Research Station
Kim Knowlton, Natural Resources Defense Council
James Lambert, Center for Risk Management of Engineering Systems / University of Virginia_
Linda Langner, US Forest Service
Jeremy Littell, University of Washington Climate Impacts Group
Herbert (Gene) Longenecker, Federal Emergency Management Agency, Region IV
George Luber, Center for Disease Control and Prevention
Julie Maldonado, National Climate Assessment / American University
Gino Marinucci, Center for Disease Control and Prevention
Mitchel McClaran, University of Arizona_
Susanne Moser, Susanne Moser Research & Consulting / Woods Institute for the Environment
Richard Moss, Pacific Northwest National Laboratory
Sheila O'Brien, National Climate Assessment_
Anthony Oliver-Smith, University of Florida
Rolf Olsen, US Army Corps of Engineers_
Adam Parris, National Oceanic and Atmospheric Administration
Jonathan Patz, University of Wisconsin
Kristina Peterson, Center for Hazards Assessment, Response and Technology, University of New Orleans
John Reilly, Massachusetts Institute of Technology
Paty Romero-Lankao, National Center for Atmospheric Research
Mike Savonis, Department of Transportation

Paul Schramm, Center for Disease Control and Prevention
Heidi Schuttenberg, National Oceanic and Atmospheric Administration_
Kathleen Sloan, Yurok Tribe
Bruce Stein, National Wildlife Federation_
Robert Thieler, US Geological Survey_
James Valverde, US Army Corps of Engineers
Thomas Webler, Social and Environmental Research Institute
Brent Yarnal, Pennsylvania State University

Stratus Consulting:
Charles Herrick, Stratus Consulting
Dave Mills, Stratus Consulting
Cameron Wobus, Stratus Consulting

Facilitators:
Mary Culver, NOAA_
David Davis, Center for Disease Control and Prevention
Dorotha Hall, Center for Disease Control and Prevention
Eric Morrisey, Center for Disease Control and Prevention
Tricia Ryan, NOAA
Lander Stoddard, Center for Disease Control and Prevention
Angela Wood, Center for Disease Control and Prevention

Appendix C: The Workshop Planning and Synthesis Team

The Workshop Planning Team
Glenis Archer, Centers for Disease Control and Prevention
Leslie Brandt, US Forest Service*
Marge Davenport, US Geological Survey
Sandy Eslinger, National Oceanic and Atmospheric Administration*
Linda Joyce, US Forest Service*
Linda Langner, US Forest Service*
James Lopez, Department of Housing and Urban Development
George Luber, Centers for Disease Control and Prevention
Julie Maldonado, National Climate Assessment / American University*
Gino Marinucci, Centers for Disease Control and Prevention
Sheila O'Brien, National Climate Assessment, USGCRP
Rolf Olsen, US Army Corps of Engineers*

* *Members of the Workshop Synthesis Team*

Appendix D: Sector-based Frameworks for VAs from the White Paper

Some of the general characteristics for framing VAs for each of the NCA sectors, based on the white paper's reviewed literature.

Application (NCA sector)	Type of data requirements	Treatment of uncertainty	References	Notes
Agriculture/ forestry	Summary of current climate variability; global climate model (GCM) output; models of crop growth; farm/ forestry decision model	Multiple climate model outputs can be compared to bracket possible outcomes; economic scenarios can be changed to represent different stakeholder responses	O'Brien et al., 2004; Littell and Peterson, 2005; Metzger et al., 2005; Schroter et al., 2005b; Spittlehouse, 2005; Berry et al., 2006; Nitschke, 2006; Johnston and Williamson, 2007; Lavorel et al., 2007; Nkem et al., 2007; Locatelli et al., 2008; Swanston et al., 2010	Agriculture/forestry vulnerability assessments can be spatially explicit, requiring some special training. They require knowledge of natural sciences (e.g., modeling future climate, crop response to drought) and economics/social sciences (e.g., farmer risk mitigation, commodity markets).
Biological diversity/ ecosystems	Information on current resources in an area, potential natural and development hazards, and possible impacts from any hazard or development are typically needed along with anticipated impacts of climate change	Quantitative and qualitative summaries of results are proposed depending on the nature of the data. Using multiple climate scenarios is also identified to address uncertainty in future projections.	Wilson et al., 2005; Manomet Center for Conservation Services and Massachusetts Division of Fisheries and Wildlife, 2010; Zack et al., 2010	Vulnerability assessments focused on biological diversity were primarily seen in the context of helping to clarify future conservation actions while accounting for climate change.
Human and social systems	Data requirements vary depending on the system of interest. May commonly include interviews with community members; socioeconomic data; and output from climate models to estimate potential exposure.	Uncertainty is not commonly explicitly evaluated.	Laidler et al., 2009; Trask, 2007; Adger et al., 2004; Ericksen, 2008; Moreno and Becken, 2009; Ford, 2006	Methodologies and data requirements for vulnerability assessments in this sector will vary significantly depending on the specific type of system being evaluated.
Human health and welfare	Summary of current climate variability; qualitative or quantitative estimates of the current burden of climate-sensitive health outcomes; qualitative or quantitative estimates of other factors that influence these outcomes	Multiple climate model outputs can be compared to bracket possible outcomes	Oliver-Smith, 1999; Brooks et al., 2005; Ebi et al., 2006; Few, 2007; Cutter and Finch, 2008; Wrachien et al., 2008; Heltberg et al., 2009; Karl et al., 2009; Strand et al., 2010; Tong et al., 2010	Health assessment models are typically constructed by individual researchers and not generally available.
Land resources	Summary of current resource status and characteristics along with relevant information from climate change scenarios	Mainly qualitative discussions in the reviewed works.	Metzger et al., 2006; Borrelli and Beavers, 2008; Metzger et al., 2008	

Application (NCA sector)	Type of data requirements	Treatment of uncertainty	References	Notes
Marine resources	Climate model output; habitat models; fisheries data (*e.g.*, catches through time); expert elicitations	Climate model uncertainty; confidence of experts in opinions	Charlotte Harbor NEP, 2009; Moreno and Becken, 2009; Grafton, 2010; McDaniels *et al.*, 2010	
Natural environment	Climate model output; habitat and physical landscape characteristics; species resilience data	Can be evaluated using results from multiple climate change scenarios; expert elicitations can be used to qualitatively rank experts' certainty	Connor and Hiroki, 2005; O' Brien *et al.*, 2006; Williams *et al.*, 2008; de Chazal *et al.*, 2008; Enquist and Gori, 2008; Preston *et al.*, 2009; Manomet Center for Conservation Services and Massachusetts Division of Fisheries and Wildlife, 2010; Romieu *et al.*, 2010; O'Leary and Galbraith, 2010; Glick and Stein, 2010	These vulnerability assessments are typically focused on particular species or habitats. Information describing habitat quality, species resilience, and potential stressors is required.
Water resources	Summary of current climate; GCM output; topography, soil, and land-use data; water supply infrastructure data; water demand and demand response model	Multiple climate model outputs can be compared to bracket possible outcomes	Jacobs *et al.*, 2005; Sullivan and Meigh, 2005; Bell *et al.*, 2008; Berkhoff, 2008; Enquist *et al.*, 2008; Obeysekera, 2008; Sharma and Barat, 2009; Wilby and Miller, 2009; Bolin *et al.*, 2010; Brown *et al.*, 2010; U.S. EPA, 2010a, 2010b	Water resource vulnerability assessments are commonly spatially explicit and quantitative; they are likely to require some special training to be implemented. These vulnerability assessments are also commonly interdisciplinary, drawing from natural and social sciences.